For EEE.

ISBN: 978-1-250-19845-7
Library of Congress Control Number: 2019932714
Our books may be purchased in bulk for promotional, educational, or business use. Please
contact your local bookseller or the Macmillan Corporate and Premium Sales Department at
(800) 221-7945 ext. 5442 or by email at MacmillanSpecialMarkets@macmillan.com.

First edition, 2019
Book design by Jennifer Browne

Printed in China by Shaoguan Fortune Creative Industries Co. Ltd., Shaoguan, Guangdong Province

1 3 5 7 9 10 8 6 4 2

The hawk family includes birds of many shapes and sizes. Here are just a few that spend time in North America.

Most hawks share five features that make them deadly hunters.

Hunting starts with watching for prey while flying high in the air, low over the ground, or waiting on a perch.

Some hawks can see ultraviolet light, which humans can't see. The urine and feces (pee and poop) some rodents mark their tracks with glow like a map showing where they are.

Once a meal is spotted, many hawks will tuck in their wings and dive or "stoop" straight down at their prey.

A GOLDEN EAGLE IN A STOOP MAY REACH UP TO 200 MILES PER HOUR!

At the last second, the hawk will open
its wings, swing its feet forward,
and grab its victim.

Bald eagles specialize in catching fish.

Golden eagles have even been seen knocking and dragging young Dall sheep off cliffs to eat on the rocks below.

Along with hunting, finding a mate
and raising babies takes up much of a
hawk's year. It all begins with
a little bit of showing off.

Courtship displays show others that a hawk is healthy, strong, and a skillful hunter.

MANY SPECIES OF HAWKS MATE FOR LIFE.

After a hawk has found a mate, it is time to build a nest. Nests are usually made of sticks and twigs and lined with soft materials like grass, leaves, or even seaweed.

The nest is made. It's ready for eggs! The female lays anywhere from one to five eggs, and the pair take turns keeping them warm.

It's like sleeping on rocks.

Red-tailed hawks

Which one are you?

GOLDEN EAGLES OFTEN LAY TWO EGGS BUT RAISE ONE CHICK. THE YOUNGER CHICK IS USUALLY EATEN BY THE OLDER.

About a month later, the chicks are hatched and hungry.

At this point, the male hawk is in charge of hunting and bringing food back to the nest.

I'll say!

Who wants fish?

We had fish for breakfast.

Bald eagles

Heads up.

BABY HAWKS OFTEN GO TO THE BATHROOM OVER THE EDGE OF THE NEST INSTEAD OF IN IT. SO TIDY!

Within a few months, the chicks have the feathers they need to fly, or "fledge." It takes some practice.

Sharp-shinned hawks

The parents will continue to feed the young birds and keep them safe as they adjust to life outside the nest.

For many hawks, winter brings a choice. Some stay put. Others fly long distances—"migrate"—to warmer climates.

Swainson's hawks will spend up to four months of the year migrating 12,000 miles in flocks called "kettles" of up to 50,000 birds. And they don't even eat for most of the trip!

They may fly fast, far, and high, but even hawks are threatened by human-caused problems on the ground.

The dangers are real, but you can begin to help by learning more about hawks and then teaching others. Because hawks are majestic animals!

WINGSPANS

Bald eagle— 6 feet

Short-tailed hawk—3 feet

Sharp-shinned hawk — 2 feet

Air movements that help hawks soar

Sun

Warm air rising

Heat from Sun

Cool forest

Hot rocks

Thermals

Wind

Mountain

Updrafts